Note on Fuzzy Clustering

Fuzzy Clustering using Levenberg-Marquardt Optimization and Deterministic Annealing

ARASH ABADPOUR

Arash Abadpour (arash@abadpour.com) is with Intellijoint Surgical Inc., Waterloo, Canada.

The author acknowledges that this text has been submitted for publication and that this file will be deleted upon publication.

Contents

1 Introduction 1

2 Literature Review 3
   2.1 Notion of Membership 3
   2.2 Prototype-based Clustering 4
   2.3 Robustification 6
   2.4 Number of Clusters 7
   2.5 Deterministic Annealing (DA) 8

3 Method 11
   3.1 Model Preliminaries 11
   3.2 Assessment of Loss 12
   3.3 Solution Strategy 14
   3.4 Outlier Detection and Classification 16
   3.5 Determination of U and λ 16
   3.6 Deterministic Annealing (DA) 18
   3.7 Implementation 20

4 Experimental Results 25

5 Conclusions 35

List of Figures

3.1 The process of selecting φ_f, based on which the value of λ is determined. 16

3.2 The process of selecting φ_p, based on which the value of U is determined. 17

3.3 The exponents that govern the ratio of $\Delta_{nc}(\psi)$ and f_{nc} values for different values of c. (a) Exponent in the ratio between $\Delta_{nc}(\psi)$ values. (b) Exponent in the ratio between f_{nc} values. 22

3.4 Impact of m on the ratio of $\Delta_{nc}(\psi)$ and f_{nc} values for different values of m. (a) Ratio between $\Delta_{nc}(\psi)$ values. (b) Ratio between f_{nc} values. 23

3.5 Impact of m on the fuzziness of the output of the clustering algorithm. (a) $m = 2$. (b) $m = 5$. 24

4.1 Inputs and outputs of the developed algorithm for a *2dl* problem instance. (a) Input data items. (b) Cluster seeds. (c) Internal variables. (d) Converged clusters. (e) Classification results. 27

4.2 Impact of the choice of m on the converged clusters for a *3dpp* problem instance. (a) $m = 2$. (b) $m = 3$. (c) $m = 4$. Input data courtesy of *Epson Edge*. 28

4.3 Impact of the choice of C on the converged clusters for a *2dc* problem instance. (a) $C = 2$. (b) $C = 3$. (c) $C = 4$. 29

4.4 Impact of the choice of φ on the converged clusters for a *2de* problem instance. (a) $\varphi = \frac{1}{4}\varphi_o$. (b) $\varphi = \varphi_o$. (c) $\varphi = 4\varphi_o$. 30

4.5 Results corresponding to the problem classes listed in Table 4.1. (a) *2dc*. (b) *2de*. (c) *2dl*. (d) *3dpp*. 31

4.6 Results corresponding to an *ics* problem instance. (a) Input data items. (b) Converged clusters. (c) Classification results. 32

4.7 Results corresponding to an *ighe* problem instance. (a) Input data items. (b) Converged clusters. (c) Classification results. 33

Abstract

In this paper, we address the generic problem of unsupervised data clustering within a fuzzy framework. We show that this problem can be formulated as a least mean square vector cost function and propose an iterative solution procedure which utilizes the Levenberg-Marquardt optimization technique. As such, we develop an algorithm which is capable of performing the stated task through utilizing abstract notions of data item, cluster, and data item-to-cluster distance. We demonstrate that the fuzzifier can be utilized as a temperature factor, in order to set up a Deterministic Annealing framework that provides assurance that the derived loss function converges to an acceptable minimum. This is important because the developed method includes an additional layer of numerical optimization on top of the structure commonly used in the literature. This paper includes the derivation of the cost function as well as the outline of the developed solution procedure. We carry sample problem instances from six problem classes and discuss different aspects of the developed method. Finally, we provide a potential next step for this work.

Chapter 1

Introduction

Unsupervised clustering of a set of data items into homogenous sets is a common requirement in many applications within fields including pattern recognition, learning, coding, and signal and image processing. While the different references to this problem utilize vastly different models for the data items and the clusters, there is a common framework in which all these problem classes are defined.

In essence, an incarnation of the unsupervised data clustering problem is based on a notion of homogeneity which is rooted in the physical properties of the data items and the clusters. Therefore, there is an important practical incentive for discussing unsupervised data clustering in generic terms and independent of the framework of any particular problem class. Under such circumstances, the vision is to develop a generic fuzzy clustering algorithm which accepts data item and cluster models as plug-ins and can operate using a data item-to-cluster distance function which is provided as a black-box.

In this paper, we demonstrate that this vision is possible and analytically derive a cost function for such a framework. We show that this formulation can be derived as a least mean square vector cost function that is suitable for the Levenberg-Marquardt optimization technique. We also show that this added level of numerical optimization requires an increase in the value of the fuzzifier and define this requirement within the context shared by previous works in the field.

We define six problem classes and demonstrate the steps required for deploying the developed algorithm within the context of these problem classes. We exhibit that this procedure is objectively planned and that there exists no need for human supervision or empirical adjustment of any parameters.

The rest of this paper is organized as follows. First, in Chapter 2 we review the literature of the subject. Then, in Chapter 3 we provide the mathematical construction of the developed method and carry an outline of the developed algorithm. This paper continues with Chapter 4, which carries experimental results, and Chapter 5, which contains the conclusions of this paper.

Chapter 2

Literature Review

2.1 Notion of Membership

The notion of membership is a key point of distinction between different clustering algorithms. Essentially, membership may be *Hard* or *Fuzzy*. K-means [1] and Hard C-means (HCM) [2] clustering algorithms, for example, utilize hard membership values. The reader is referred to [3] and the references therein for a history of K-means clustering and other methods closely related to it. Iterative Self-Organizing Data Clustering (ISODATA) [4] is a hard clustering algorithm as well.

With the introduction of Fuzzy Theory [5], many researchers incorporated this more natural notion into clustering algorithms [6, 7]. In the words of the authors of [8], a fuzzy membership regime is more applicable where "a more nuanced description of the objects affinity to the specific cluster is required". Additionally, from a practical perspective, it is observed that hard clustering techniques are extremely more prone to falling into local minima [9]. The reader is referred to [10] for a recent review of the field of fuzzy clustering.

Initial work on fuzzy clustering was done by Ruspini [11, 12] and Dunn [13, 14] and it was then generalized by Bezdek [15, 16] into Fuzzy C-means (FCM). In FCM, data items, which are denoted as $x_1, \cdots, x_N$, belong to $\mathbb{R}^k$ and clusters, which are identified as $\psi_1, \cdots, \psi_C$, are represented as points in $\mathbb{R}^k$. FCM makes the assumption that the number of clusters, C, is known through a separate process or expert opinion and minimizes the following objective function,

$$\Delta = \sum_{c=1}^{C} \sum_{n=1}^{N} f_{nc}^m \|x_n - \psi_c\|^2, \qquad (2.1)$$

subject to,

$$\sum_{c=1}^{C} f_{nc} = 1, \forall n. \qquad (2.2)$$

Here, $f_{nc} \in [0,1]$ denotes the membership of x_n to ψ_c.

In (2.1), $m > 1$ is the *fuzzifier* (also called *weighing exponent* and *fuzziness*). The optimal choice for the value of the fuzzifier is a debated matter [17] and is suggested to be "an open question" [18]. Bezdek [19] suggests that $1 < m < 5$ is a proper range and utilizes $m = 2$. The use of $m = 2$ is suggested by Dunn [13] in his early work on the topic as well and also by Frigui et al. [20], among others [21]. Bezdek [22] provided physical evidence for the choice of $m = 2$ and Pal et al. [23] suggested that the best choice for m is probably in the interval $[1.5, 2.5]$. Yu et al. [18] argue that the choices for the value of m are mainly empirical and lack a theoretical basis. Nevertheless, it is known that larger values of m soften the boundary between the clusters [24]. The reader is referred to [25] for a review of the concept of fuzzifier and the alternatives for it.

FCM is the groundbreaking mathematical model and the gateway into a majority of other algorithms in the literature. In other words, a significant number of subsequent works in the field modify FCM in order to alter its behavior and to achieve desired properties. Augmentation of FCM with regularization terms and constraints, however, is not without its inherent hazards. In fact, many works in the literature employ such techniques and as a by-product become dependent on additional parameters which need to be set meticulously. The authors of [26] provide a list of some of the affected approaches.

2.2 Prototype-based Clustering

It is a common assumption that the notion of homogeneity depends on the distances between the data items. This assumption is made implicitly when clusters are modeled as *prototypical* data items, also called *clustroids* or cluster *centroids*, as in FCM, for example. A prominent choice in these works is the use of the Euclidean distance function [27]. For example, the potential function approach considers data items as energy sources scattered in a multi-dimensional space and seeks peak values in the field [28]. We argue, however, that the *distance* between the data items may not be either defined or meaningful and that what the clustering algorithm is to accomplish is the minimization of *data item-to-cluster* distances. For example, when data items are to be clustered into certain lower-dimensional subspaces, as it is the case with Fuzzy C-Varieties (FCV) [29], the

Euclidean distance between the data items is irrelevant. We note that, in fact, fuzzy clustering is sometimes equated and reduced to prototype-based clustering [24] (this reductive perspective is prevalent as of 2015 [30]).

Prototype-based clustering does not necessarily require prototypes which are explicitly present. For example, in kernel-based clustering [31], it is assumed that a non-Euclidean distance can be defined between any two data items. The clustering algorithm then functions based on an FCM-style objective function and produces clustroids which are defined in the same feature space as the data items [32]. These cluster prototypes may not be explicitly represented in the data item space, but, nevertheless, they share the same mathematical model as the data items [33]. Fuzzy Analysis (FANNY) [34] is another algorithm in which, although there are no prototypes, but, nevertheless, homogeneity is based on the mutual distances between the data items.

Relational clustering approaches constitute another class of algorithms which are intrinsically based on the distances between the data items (for example refer to Relational FCM (RFCM) [35] and its non-Euclidean extension Nerf C-means [36]). The goal of this class of algorithms is to group the data items into *self-similar* bunches. Fuzzy clustering by Local Approximation of Memberships (FLAME) [37] and Hierarchical Agglomerative Clustering (HAC) [38, 14.3.12 Hierarchical clustering] are other clustering algorithms which inherently guide the process of clustering based on the distances between the data items. Additionally, some researchers utilize ℓ_r-norms, for $r \neq 2$ [39, 40, 41], or other distance functions which are defined between a pair of data items [42].

We argue that a successful departure from the assumption of prototypical clustering is achieved when clusters and data items have different mathematical models. For example, the Gustafson-Kessel algorithm [43] models a cluster as a pair of a point and a covariance matrix and utilizes the Mahalanobis distance between data items and clusters (also see the Gath-Geva algorithm [44] and fuzzy c-regressions [45]. Fuzzy shell clustering algorithms [21], which are sometimes addressed as Fuzzy C-Shells (FCS), utilize more generic geometrical structures. For example, the FCV [29] algorithm can detect lines, planes, and other hyper-planar forms, the Fuzzy C Ellipsoidal Shells (FCES) [46] algorithm searches for ellipses, ellipsoids, and hyperellipsoids, and the Fuzzy C Quadric Shells (FCQS) [21] and its variants seek quadric and hyperquadric clusters.

2.3 Robustification

Dave and Krishnapuram [47] argue that the function of membership values in FCM and the concept of weight functions in robust statistics are related. Based on this perspective, it is argued that the classical FCM in fact provides an indirect means for attempting robustness. Nevertheless, it is known that FCM and other least square methods are highly sensitive to noise [48]. The authors of [49] in fact perform a numerical review of this topic in the context of FCM and the Possibilistic C-means (PCM) and argue that the FCM constraints are "too strong" and that PCM constraints, on the other hand, are "too weak". Hence, there has been ongoing research on the possible modifications of FCM in order to provide a (more) robust clustering algorithm [50, 51]. Dave et al. [47] provide an extensive list of relevant works and outline the intrinsic similarities within a unified view (also see [52, 53]).

The first attempt to robustifying FCM, based on one account [47], is the Ohashi Algorithm [52, 54] which adds a noise cluster to FCM. That transformation was suggested independently by Dave [53] when he developed the Noise Clustering (NC) algorithm as well (also see Robust Fuzzy Clustering Algorithm (RFCA) [55]). The core idea in NC is that there exists one additional imaginary prototype which is at a fixed distance from all of the data items and represents noise.

Krishnapuram et al. [56] extended the idea behind NC and developed the PCM algorithm by rewriting the objective function as,

$$\Delta = \sum_{c=1}^{C}\sum_{n=1}^{N} t_{nc}^m \|x_n - \psi_c\|^2 + \sum_{c=1}^{C} \eta_c \sum_{n=1}^{N}(1 - t_{nc})^m. \tag{2.3}$$

Here, t_{nc} denotes the degree of representativeness or *typicality* of x_n to ψ_c (also addressed as a *possibilistic degree* in contrast to the *probabilistic* model utilized in FCM).

Borrowing the language used in [27], in FCM, clusters "seize" data items and it is disadvantageous for multiple clusters to claim high membership to the same data item. There is no phenomenon, however, in NC and PCM which corresponds to this internal factor. Additionally, it is likely that PCM clusters coincide and/or leave out portions of the data unclustered [57]. In fact, it is argued that the fact that at least some of the clusters generated through PCM are non-coincidental is because PCM gets trapped into local minimum [58] (also see [25, 24]). PCM is also known to be more sensitive to initialization and the exact values of the configuration parameters compared to the other algorithms in its class [57, 27].

It has been argued that both concepts of possibilistic degrees and membership values have positive contributions to the purpose of clustering [59, 60]. Hence, Pal et al. [59] combined FCM and PCM and rewrote the optimization function of Fuzzy Possibilistic C-Means (FPCM). That approach was later shown to suffer from different scales for the membership terms, especially when $N \gg C$, and, therefore, additional linear coefficients and a PCM-style term were introduced to the objective function [61]. It has been argued that the resulting objective function employs four correlated parameters and that the optimal choice for them for a particular problem instance may not be trivial [27]. Additionally, in the new combined form, f_{nc} cannot necessarily be interpreted as a membership value [27]. The reader is also referred to Possibilistic-Fuzzy C-Means (PFCM) [49] for a related model which "hybridizes" PCM and FCM in order to "avoid various problems of PCM, FCM, and FPCM". Additionally, see [62] for other variants and [63] for an added entropy-style regularization term.

The introduction of the PCM model is motivated by several factors, amongst which is to be able to relax, or somewhat circumvent, the sum-of-one constraint for the membership values. As such, through "giving up the requirement for strict partitioning" [64], the expectation is that the resulting algorithm will be able to reject outliers and to deal with data items which do not belong to any of the clusters more efficiently. As discussed here, however, the utilization of PCM-style models has given rise to the emergence of other difficulties. In this context, the model developed in [65] is worth particular attention. That work states that the relationship between the data items and the clusters must be assessed at two levels, i.e. whether or not a data item is an outlier and, if not, which clusters it belongs to. In other words, the model developed in [65] replaces the singleton membership identifiers f_{nc} with the pair of p_n and f_{nc}. Here, p_n models the probability that x_n is an inlier and f_{nc} models the probability that it belongs to ψ_c, given that it is an inlier. An extension of this model in [66] demonstrates that the p_n variables can emerge from the f_{nc} values when the classical parallel clustering framework is substituted with a serially-structured pipeline.

Frigui et al. [20] included the robust loss function $u_c(\cdot)$ in the objective function of FCM and developed Robust C-Prototypes (RCP). They further extended RCP and developed an unsupervised version of RCP, nicknamed URCP [20]. Wu et al. [33] used $u_c(x) = 1 - e^{-\beta x^2}$ and developed Alternative HCM (AHCM) and Alternative FCM (AFCM).

2.4 Number of Clusters

The classical FCM and PCM, and many of their variants, are based on the assumption that the number of clusters is known (an extensive review of this topic is given in [16, Chapter 4]). While

PCM-style formulations may appear to relax this requirement, the corresponding modification is carried out at the cost of yielding an ill-posed optimization problem [27]. In fact, repeating the clustering procedure for different numbers of clusters [44, 67] and *Progressive Clustering* are two of the alternative approaches to address the challenge of not requiring *a priori* knowledge about the number of clusters present in a particular data.

Among the many variants of Progressive Clustering are methods which start with a significantly large number of clusters and freeze "good" clusters [68], approaches which combine compatible clusters [68, 20], and the technique of searching for one "good" cluster at a time until no more is found [69]. These approaches utilize one or more Cluster Validity Indexes (CVI) [70, 71]. Use of regularization terms in order to push the clustering results towards the "appropriate" number of clusters is another approach taken [59].

Dave and Krishnapuram conclude in their 1997 paper that the solution to the general problem of robust clustering when the number of clusters is unknown is "elusive" and that the techniques available in the literature each have their limitations [47]. In this paper, we acknowledge that the problem of determining the appropriate number of clusters is hard to solve and even hard to formalize. Therefore, we designate this challenge as being outside the scope of this contribution and assume that the appropriate number of clusters is known *a priori*. The reader is referred to the category of methods collectively known as Visual Assessment of cluster Tendency (VAT) [72] in this context. More recent variants of VAT include Automated VAT (aVAT) [73] and Improved VAT (iVAT) [74].

2.5 Deterministic Annealing (DA)

A thorough review of the concept of Annealing, including Simulated and Deterministic Annealing (DA), is outside the scope of this paper. In summary, it is observed that certain chemical systems can be driven to their lower-energy state by annealing. During this process, the system is transitioned from a high-temperature state to one in which the system temperature is significantly lower. Physical observation confirms that as the temperature is lowered, the system becomes more discriminating, i.e. the probable state-space of the system shrinks in volume. Simulated Annealing employs this physical analogy in order to solve optimization problems which involve non-convex cost functions [75, 76].

Deterministic Annealing (DA) is the deterministic counterpart to Simulated Annealing, in that it utilizes the same concept of a system moving towards more rigid settings through lowering a temperature variable. Nevertheless, DA avoids a statistical approach, but maintains a model in which local minimums are suppressed at higher temperatures. An analogy is to consider DA as a Simulated Annealing system in which stochastic simulation is replaced with expectations [77]. In practice, the term DA is often liberally used in order to address any system which is parametrized by a pseudo temperature, the use of which provides assurances for avoiding local minimums [78, 79, 80].

Within the context of clustering, the primary approach to utilizing DA is to employ a temperature term which governs the magnitude of the randomness of the solution to the clustering problem [77]. In this approach, which is exemplified in [81, 82, 83], the solution moves from total ambiguity, in which case every data item belongs equally to every cluster, to a hard solution in which binary membership is achieved. The actual implementation of this idea involves the employment of the temperature term as the regularization coefficient which governs the influence of an entropy term. Hence, when the system starts at infinite temperature, it essentially maximizes the entropy, while at lower temperatures, the loss associated to the clustering solution is also gradually taken into consideration.

Chapter 3

Method

3.1 Model Preliminaries

We assume that any problem class provides a mathematical model for the data items and denote a data item as x. We also assume that any problem class defines a cluster model which complies with the *notion of homogeneity* relevant to the problem class at hand, and denote a cluster as ψ.

In this work, we utilize a weighted set of data items, defined as,

$$\mathbf{X} = \left\{ (\omega_n; x_n) \right\}, n = 1, \cdots, N, \omega_n > 0, \qquad (3.1)$$

and we define the *weight* of $\mathbf{X}$ as $\Omega(\mathbf{X}) = \sum_{n=1}^{N} \omega_n$. When known in the context, we abbreviate $\Omega(\mathbf{X})$ as Ω. Thus, when estimating expected values, we treat $\mathbf{X}$ as a set of realizations of the random variable x and write,

$$p\{x_n\} = \frac{\omega_n}{\Omega}. \qquad (3.2)$$

Note that a problem class generally relates to a physical phenomenon which mandates a specific relationship between the data items and the clusters. In this work, we model this relationship as the real-valued positive *distance function* $\phi(x, \psi)$. Through this abstraction, we decidedly avoid the dependence of the underlying algorithm on Euclidean or any other particular notations of distance. We also assume that the distance function is unbounded, i.e. that for any cluster representation ψ and any positive value L, there exist infinite number of data items x for which $\phi(x, \psi) > L$.

We also assume that a function $\Psi_\circ(\cdot)$, which may depend on $\mathbf{X}$, is given, that produces a relevant set of initial clusters. We address this entity as the *cluster initialization function* and denote the number of clusters produced by it as C.

We assume that the robust loss function, $u(\cdot) : [0, \infty] \to [0, 1]$, is given which satisfies $\lim_{\tau \to \infty} u(\tau) = 1$. Additionally, we assume that $u(\cdot)$ is an increasing differentiable function which satisfies $u(0) = 0$ and $u(1) = \frac{1}{2}$.

In this work, we utilize the rational robust loss function,

$$u(x) = \frac{x}{1+x}, \qquad (3.3)$$

and we model the loss of x_n when it belongs to ψ_c as,

$$u_{nc} = u\left(\frac{1}{\lambda}\phi_{nc}\right), \phi_{nc} = \phi(x_n, \psi_c). \qquad (3.4)$$

Additionally, we model the loss of a data item which is considered to be an outlier as the positive constant U.

We address λ as the *scale* parameter (note the similarity with the cluster-specific weights in PCM [56]). In fact, λ has a similar role to that of scale in robust statistics (also called the *resolution parameter* [84]) and the idea of distance to noise prototype in the NC algorithm [53, 55]. Scale can also be considered as the controller of the boundary between inliers and outliers [47]. From a geometrical perspective, λ controls the radius of spherical clusters and the thickness of planar and shell clusters [27].

This modeling framework has precedence in the literature [65].

3.2 Assessment of Loss

We assume that, at some point during the operation, the developed clustering procedure has found the C clusters $\psi_1, \cdots, \psi_C$. We also assume that a Maximum Likelihood procedure has been applied on $\mathbf{X}$ and denote the set of data items which are assigned to ψ_c as $\tilde{\mathbf{X}}_c$. We address the union of all $\tilde{\mathbf{X}}_c$ for $c = 1, \cdots, C$ as $\tilde{\mathbf{X}}$. In this context, the set $\tilde{\mathbf{X}}_0 = \mathbf{X} - \tilde{\mathbf{X}}$ contains the data items which are considered to be outliers.

Now, we consider an arbitrary data item x_n. This data item may be an outlier or it may belong to one of the C clusters. Hence, we model the loss associated with x_n as follow.

$$E\{Loss|x_n\} = p\left\{x_n \in \tilde{\mathbf{X}}_0\right\} E\left\{Loss|x_n \in \tilde{\mathbf{X}}_0\right\} \qquad (3.5)$$
$$+ p\left\{x_n \in \tilde{\mathbf{X}}\right\} \sum_{c=1}^{C} p\left\{x_n \in \tilde{\mathbf{X}}_c | x_n \in \tilde{\mathbf{X}}\right\}$$
$$E\left\{Loss|x_n \in \tilde{\mathbf{X}}_c\right\}.$$

We now denote the probability that x_n is an inlier as p_n and the probability that x_n belongs to $\tilde{\mathbf{X}}_c$, given that it is an inlier, as f_{nc} and rewrite (3.5) as follows. Here, we also used (3.4).

$$E\{Loss|x_n\} = (1-p_n)U + p_n \sum_{c=1}^{C} f_{nc} u_{nc}. \tag{3.6}$$

Hence,

$$E\{Loss|\mathbf{X}\} = \sum_{n=1}^{N} p\{x_n\} E\{Loss|x_n\} = \tag{3.7}$$

$$\frac{1}{\Omega} \sum_{n=1}^{N} \omega_n \left[p_n \sum_{c=1}^{C} f_{nc} u_{nc} + UC \frac{1}{C}(1-p_n) \right].$$

Close assessment of (3.7) shows that this cost function complies with an HCM-style hard template. It is known, however, that the utilization of the concept of the fuzzifier has important benefits, as outlined in Section 2.1. Hence, we rewrite (3.7) and derive the following objective function for the clustering problem proposed in this paper.

$$\Delta = \sum_{n=1}^{N} \omega_n \left[p_n^m \sum_{c=1}^{C} f_{nc}^m u_{nc} + UC^{1-m}(1-p_n)^m \right]. \tag{3.8}$$

This objective function is to be minimized subject to (2.2). Note that, compared to [65], the present work utilizes a general m. As we will show later, this is an important property of the present work and allows for more optimal mitigation of local minimums in the context of the present formulation.

In (3.7) we have chosen to write U as $UC\frac{1}{C}$ in order to compensate for the impact of the fuzzifier. In fact, two types of terms exist in (3.7), i.e. $p_n f_{nc}$ and $(1-p_n)$. These terms are each products of membership identifiers. However, while the first term contains two elements, the second one only contains the single element p_n. We argue that this is because this term in fact contains implicit components of the type *"if either P or not P"* hidden in it. In other words, the cost component $U(1-p_n)$ in fact models the situation in which x_n is an outlier, in which case it is irrelevant whether or not x_n belongs to any of the clusters. In other words, the term $(1-p_n)$ is in fact the simplified version of the following term,

$$(1-p_n) \sum_{c=1}^{C} f_{nc} = 1 - p_n. \tag{3.9}$$

While this alternative form is in effect identical to $1-p_n$, the difference becomes significant when the fuzzifier is integrated into the objective function. In fact, with the addition of the fuzzifier, the term given in (3.9) ought to be modified to,

$$(1-p_n)^m \sum_{c=1}^{C} f_{nc}^m \leq (1-p_n)^m. \tag{3.10}$$

In other words, if no other measure is taken, the incorporation of the fuzzifier will effectively reduce the cost of being an outlier, as explained below.

We note that for any set of K non-negative variables ζ_k which satisfy $\sum_{k=1}^{K} \zeta_k = 1$, we have $\sum_{k=1}^{K} \zeta_k^m \leq 1$, when $m > 1$. The equality in this relationship, i.e. the upper bound, occurs when all of the ζ_k are zero except for one which is unity. The lower bound on $\sum_{k=1}^{K} \zeta_k^m$, however, occurs when the ζ_k are identical. Therefore, we replace (3.9) with the case in which all the f_{nc} are equal. This process guarantees that when the fuzzifier is incorporated into the cost function, the corresponding term is always greater than or equal to the pre-fuzzifier term. In other words, we replace $(1-p_n)$ with $(1-p_n)C\frac{1}{C}$ and therefore, after the incorporation of the fuzzifier, yield $(1-p_n)^m C \frac{1}{C^m} = (1-p_n)^m C^{1-m}$. In the above, this transformation was rephrased, *imprecisely*, as substituting U with $UC\frac{1}{C}$ in (3.7).

3.3 Solution Strategy

First, we optimize (3.8) subject to f_{nc}. To do so, we utilize Lagrange Multipliers and incorporate (2.2) into (3.8) and derive,

$$f_{nc} = \frac{u_{nc}^{-\frac{1}{m-1}}}{\sum_{c'=1}^{C} u_{nc'}^{-\frac{1}{m-1}}}. \tag{3.11}$$

We then calculate the partial derivative of Δ relative to p_n and use (3.11) and derive,

$$p_n = \frac{\sum_{c=1}^{C} u_{nc}^{-\frac{1}{m-1}}}{CU^{-\frac{1}{m-1}} + \sum_{c=1}^{C} u_{nc}^{-\frac{1}{m-1}}}. \tag{3.12}$$

Note that both (3.11) and (3.12) only depend on u_{nc}, and not on f_{nc} or p_n.

We now utilize (3.11) and (3.12) and rewrite (3.8) as,

$$\Delta = \Delta_0 + \sum_{c=1}^{C} \left\| \vec{\Delta}_c (\psi_c) \right\|^2. \tag{3.13}$$

Here,

$$\Delta_0 = \sum_{n=1}^{N} w_n U C^{1-m} (1-p_n)^m, \tag{3.14}$$

$$\vec{\Delta}_c(\psi) = \begin{bmatrix} \Delta_{1c}(\psi) \\ \vdots \\ \Delta_{Nc}(\psi) \end{bmatrix}, \tag{3.15}$$

$$\Delta_{nc}(\psi) = \left(\frac{u_{nc}^{-\frac{1}{m-1}}}{CU^{-\frac{1}{m-1}} + \sum_{c'=1}^{C} u_{nc'}^{-\frac{1}{m-1}}} \right)^{\frac{m}{2}} \sqrt{\omega_n u \left(\frac{1}{\lambda} \phi(x_n, \psi) \right)}. \tag{3.16}$$

Note that, Δ_0 is independent of any ψ_c and also that $\vec{\Delta}_c(\psi_c)$ is independent of any $\psi_{c'}$ for $c \neq c'$. Hence, in order to minimize Δ in terms of ψ_c, it suffices to set ψ_c to the ψ which minimizes $\|\vec{\Delta}_c(\psi)\|^2$. Assessment of (3.15), however, shows that this task can be accomplished using the Levenberg–Marquardt algorithm [85] (also see [86, 87]).

Note that, although the current work utilizes the modeling framework developed previously [65], it does not require that the new ψ_c is examined in order to confirm that it reduces the value of the cost function. This is one of the contributions of the present work and allows the developed algorithm to avoid an expensive step for every cluster and every iteration. Additionally, the present work eliminates the need for the calculation of p_n and f_{nc} during the iterations. These identifiers, in fact, can be calculated after convergence is achieved, if the user desires so. In fact, as will be shown later, the developed method does not require that the values of f_{nc} and p_n are ever calculated.

Nevertheless, the developed method requires that Δ is calculated and monitored in order to determine whether the algorithm has converged. Hence, we substitute (3.11) and (3.12) in (3.8) and show that,

$$\Delta = \sum_{n=1}^{N} \frac{\omega_n}{\left(CU^{-\frac{1}{m-1}} + \sum_{c=1}^{C} u_{nc}^{-\frac{1}{m-1}} \right)^{m-1}} \tag{3.17}$$

Note that, the framework developed in this paper eliminates the need for the *cluster fitting function*, i.e. $\Psi(\cdot)$, as it is required in [65]. In fact, through utilizing the Levenberg–Marquardt algorithm, the developed algorithm produces a procedural alternative for $\Psi(\cdot)$. This modification, as will be explained later, is only possible because the method developed in this paper utilizes a DA framework.

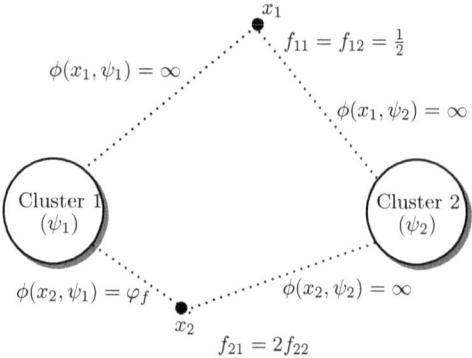

Figure 3.1: The process of selecting φ_f, based on which the value of λ is determined.

3.4 Outlier Detection and Classification

The relationship between x_n and ψ_c is affected by both f_{nc} and p_n. We emphasize, however, that this situation is characteristically different from FPCM, where a clear notion of membership to cluster cannot be inferred from the variables involved in the model [27].

Taking the arbitrary data item x_n, we recognize that p_n denotes the probability that x_n is an inlier. Hence, if $p_n < \frac{1}{2}$, we add x_n to the set of outliers. Otherwise, we include x_n in $\tilde{\mathbf{X}}_{c_n}$. Here, the integer $1 \leq c_n \leq C$ denotes the cluster for which f_{nc} is the largest (i.e. $c_n = \arg_c \max\{f_{nc}\}$). For the purpose of coherence, we define $c_n = 0$ for outliers.

Review of (3.11) and (3.12) shows that the calculation of f_{nc} and p_n is not a requirement for the calculation of c_n. In fact, direct derivation shows that,

$$c_n = \begin{cases} 0 & \sum_{c=1}^{C} u_{nc}^{-\frac{1}{m-1}} < CU^{-\frac{1}{m-1}} \\ \arg\min_c u_{nc} & \text{otherwise} \end{cases} \quad (3.18)$$

3.5 Determination of U and λ

It is evident that λ defines the scale for ϕ_{nc}. This is exemplified in (3.4) and also everywhere else in this paper where ϕ_{nc} is divided by λ. We argue that, similarly, U defines the scale for u_{nc}. Hence, we argue that the two identities ϕ_{nc} and u_{nc} are *brought into context* through λ and U, respectively. We use this perceptual definition in order to propose a procedure for determining the appropriate values for λ and U for a particular problem class.

We suggest an imaginary situation, as depicted in Figure 3.1, in which two data items interact

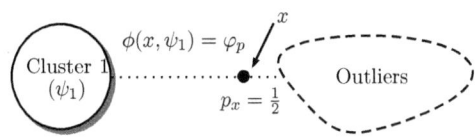

Figure 3.2: The process of selecting φ_p, based on which the value of U is determined.

with two clusters ($N = 2$ and $C = 2$). The first data item, here x_1, is infinitely far from both clusters, in which case we expect $f_{11} = f_{12} = \frac{1}{2}$. This equilibrium is in fact verified by (3.11). The second data item, here x_2, however, is infinitely far from the second cluster and is at the distance of φ_f from the first one. Here, we ask what value of φ_f will result in $f_{21} = 2f_{22}$. This question can be asked in a different setting as follows: *For a data item which is infinitely far from two clusters, how close should it get to one cluster, while maintaining its distance to the other, in order for it to be favored by the former cluster two times than the latter?* Utilizing (3.11) one can show that,

$$\lambda = \varphi_f \left(2^{m-1} - 1 \right). \tag{3.19}$$

In order to estimate the proper value for U, we utilize another imaginary situation in which one data item interacts with one cluster, as depicted in Figure 3.2. Here, we ask how far the data item should be from the cluster in order for it to be an outlier with a probability of half? This situation, in effect, defines the boundary of inliers and outliers when Maximum Likelihood is applied on p_n. We denote this distance as φ_p and derivation using (3.12) shows that,

$$U = u \left(\frac{1}{\lambda} \varphi_p \right). \tag{3.20}$$

It is important to emphasize that the outcome of this process is independent of C. In other words, in a set of data items which contains C clusters, a data item which is at the same distance of φ_p from all of them will be an inlier with the probability of half. This is the result of the decision to replace U with $UC\frac{1}{C}$ in (3.7). In fact, in the absence of this substitution, (3.20) changes to the following equation,

$$U = \frac{1}{C^{m-1}} u \left(\frac{1}{\lambda} \varphi_p \right). \tag{3.21}$$

In this equation, U depends on C. In other words, the cost associated to an outlier depends on the number of clusters present in the system. We find this dependence hard to understand from a perceptual standpoint. The utilized formulation, i.e. the one given in (3.20), however, yields the case in which U is a property of the problem class and addition or removal of clusters does not affect it.

We note that φ_p and φ_f reflect different aspects of the relationship between a data item and a cluster. Nevertheless, conceptually, we can envision that one may want to set $\varphi_p = \varphi_f = \varphi$. In this line of reasoning, φ denotes the territory of a cluster, within which the cluster considers a data item as an inlier, therefore $p_x \geq \frac{1}{2}$, and also owns the data item when in competition with another farther cluster, therefore $f_{xc} \geq \frac{1}{2}$. Reworking (3.20) for $\varphi_p = \varphi_f = \varphi$ we arrive at $U = 2^{-(m-1)}$.

3.6 Deterministic Annealing (DA)

The derivations carried in Section 3.3 do not specify a particular value of m. This general approach needs to be considered in the context of the discussion given in Section 2.1, therein, one of the key differences between FCM and HCM is the introduction of the concept of fuzzifier. This distinct property of FCM was later inherited by the category of methods which grew out of FCM as well.

We note that setting $u(x) \equiv x$ and $U = \infty$ converts (3.8) into the FCM model. Moreover, the additional assumption of $m = 1$ converts this model into HCM. Hence, in essence, the model developed in this paper is a generalization of the both HCM and FCM. In this context, one may use (3.16) and (3.11) and write,

$$\frac{\Delta_{nc_1}(\psi)}{\Delta_{nc_2}(\psi)} = \left(\frac{u_{nc_1}}{u_{nc_2}}\right)^{-\frac{m}{2(m-1)}}, \qquad (3.22)$$

$$\frac{f_{nc_1}}{f_{nc_2}} = \left(\frac{u_{nc_1}}{u_{nc_2}}\right)^{-\frac{1}{m-1}}. \qquad (3.23)$$

Hence, it is the value of the exponents in (3.22) and (3.23) which control the impact of u_{nc} on cluster representations as well data item-to-cluster membership. Here, we review these exponents.

Figure 3.3 provides a pictorial representation of the ratio between $\Delta_{nc}(\psi)$ and f_{nc} pairs for different values of m. Here, HCM and FCM are represented as the cases in which $m \to 1$ and $m = 2$, respectively. In fact, a more accurate representation is to identify FCM as any point on the m axis, which, nevertheless, is constant.

Note that, both of the curves shown in Figure 3.3 tend to $-\infty$ for HCM. This can be interpreted as a winner-take-all situation in which the u_{nc} which is slightly smaller pushes the significance of the others to zero. As a result, clusters are crisp, because x_n has no impact on ψ_c, if there exists a c^* for which $u_{nc^*} < u_{nc}$. Moreover, $f_{nc} = 0$ if there exists a c^* for which $u_{nc^*} < u_{nc}$.

FCM alleviates this situation by utilizing a finite value for both exponents and experimental results collected in the past 40 years show that this transformation leads to more stable algorithms. The present work adopts this transformation and extends it. The purpose of this extension is, first, to reduce sensitivity to local minimums, and, two, to limit fuzziness of the results. It is worth emphasizing that while HCM and FCM operate on a constant-m platform, the developed work in fact utilizes the thick gray lines in Figure 3.3. In other words, the present work adopts the departure of FCM from HCM, in terms of the value of m, and thus utilizes a large value of m, to the extent necessary, and then incrementally reduces m in order to reach to the $m = 2$ situation, which yields desirable limits on the fuzzyness of the system. These concepts are discussed next.

Figure 3.4 shows the curves corresponding to $\frac{\Delta_{nc_1}(\psi)}{\Delta_{nc_2}(\psi)}$ and $\frac{f_{nc_1}}{f_{nc_2}}$ for different values of m. These curves are, in effect, manifestations of (3.22) and (3.23). Note that, we have excluded the case of $m = 1$, which yields discontinuous curves which produce 0, 1, and ∞. Note that, all the curves shown in Figure 3.4 pass through the $[1, 1]$ point, as expected. These curves, however, show the variation in the rate of change of the ratio between the corresponding pair as the value of $\frac{u_{nc_1}}{u_{nc_2}}$ moves away from one in either direction.

In fact, we notice that both sets of curves flatten as m grows. In other words, larger values of m diminish the relative significance of more likely members of a cluster in terms of their impact on the ratio between both $\Delta_{nc}(\psi)$ values as well as f_{nc} values. This situation is, in effect, a more fuzzy system, in which the distinction between different data items is weakened. Here, we provide an example.

Figure 3.5 shows two sample results corresponding to the *ighe* problem class (refer to Table 4.1 for details about this and other problem classes discussed in this paper). In these figures, the thick dashed line indicates the histogram of the input image, and therefore ω_n, whereas the thin dashed line shows p_n. Colored lines in these graphs show values of f_{nc}. Aside from the other differences between the two graphs, they demonstrate that larger values of m reduce the sharpness of the membership curves. This observation is in compliance with the findings cited for Figure 3.4.

Hence, on the surface, there is an inherent paradox present in the concept of fuzziness. On the one hand, smaller m is known to correspond to higher chances of local minimum difficulties and therefore, as shown in Figure 3.3, the FCM family increases m in order to "soften" the cost manifold and to achieve a more optimal convergence. On the other hand, as shown in Figure 3.4, larger m increases the fuzziness of the system and increases the vagueness of the classification results. It appears as if one would like to increase m in order to bypass local minimums and, at the same time, adopt a small m in order to limit vagueness. In this context, "small m" may refer to $m = 2$ which is preferred both theoretically [22] and empirically.

We propose to avoid picking a side in this apparent paradox and to break the dichotomy of small m vs. large m using a dynamic m, as follows. To do so, the developed method utilizes a large value of m which "softens" the system into a "smoother manifold" (note the intentional usage of temperature-related annealing metaphors). Then, after the system settles into a stable state, we drop m and allow the system to move to a more optimal state in this "colder manifold" in which vagueness is reduced and hence unambiguous classification is possible. We repeat this procedure until m reaches 2, the preferred value.

This process is pictorially represented in Figure 3.3, where the proposed method utilizes an interval of m values, as opposed to the individual points utilized by HCM, FCM, and a significant majority of the works present in the literature.

3.7 Implementation

We utilize the common entities which are present in the mathematical formulas carried in Section 3.3 and define,

$$u_n = CU^{-\frac{1}{m-1}} + \sum_{c=1}^{C} u_{nc}^{-\frac{1}{m-1}}. \tag{3.24}$$

Now, (3.16) simplifies to,

$$\Delta_{nc}(\psi) = \frac{u_{nc}^{-\frac{m}{2(m-1)}}}{u_n^{\frac{m}{2}}} \sqrt{\omega_n u \left(\frac{1}{\lambda}\phi(x_n, \psi)\right)}, \tag{3.25}$$

and (3.17) is rewritten as,

$$\Delta = \sum_{n=1}^{N} \frac{\omega_n}{u_n^{m-1}}. \tag{3.26}$$

Additionally, (3.18) is simplified to,

$$c_n = \begin{cases} 0 & u_n < 2CU^{-\frac{1}{m-1}} \\ \arg\min_c u_{nc} & \text{otherwise} \end{cases}, \tag{3.27}$$

and p_n and f_{nc} are rewritten as follows,

$$p_n = 1 - \frac{CU^{-\frac{1}{m-1}}}{u_n}, \tag{3.28}$$

$$f_{nc} = \frac{u_{nc}^{-\frac{1}{m-1}}}{u_n - CU^{-\frac{1}{m-1}}}. \tag{3.29}$$

Here, we utilize these alternative forms and summarize the method developed in this paper as the following Picard iteration.

(a) Call $\Psi_o\left(\cdot\right)$ and produce $\psi_1, \cdots, \psi_C$.

(b) Loop

  (a) Calculate ϕ_{nc} and u_{nc} for all c and n using (3.4).

  (b) Calculate u_n for all n using (3.24).

  (c) Produce $\vec{\Delta}_c\left(\psi\right)$ for all c using (3.25) and (3.15).

  (d) Calculate $\psi_c = \arg\min_\psi \|\vec{\Delta}_c\left(\psi\right)\|^2$ for all c.

  (e) Calculate Δ using (3.26).

  (f) If change in Δ is negligible

    i. If $m = 2$, break the loop.

    ii. Decrement m.

(c) Perform classification for all n using (3.27), if required.

(d) Calculate p_n and f_{nc} for all n and c using (3.28) and (3.29), if required.

Note that this outline assumes that m is an integer. This assumption can be alleviated with careful examination of m when it is decremented. Also, note that the calculations given in Section 3.5, for producing λ and U, are m-dependent and that therefore they need to be redone when m changes.

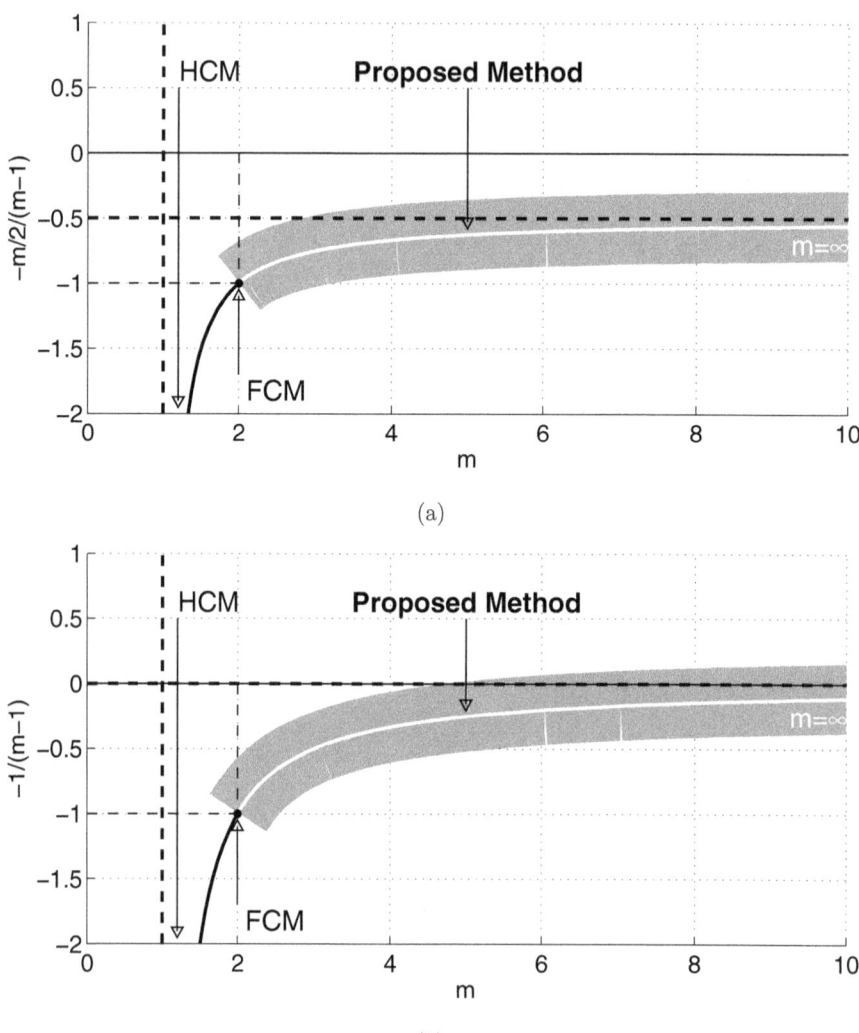

Figure 3.3: The exponents that govern the ratio of $\Delta_{nc}(\psi)$ and f_{nc} values for different values of c. (a) Exponent in the ratio between $\Delta_{nc}(\psi)$ values. (b) Exponent in the ratio between f_{nc} values.

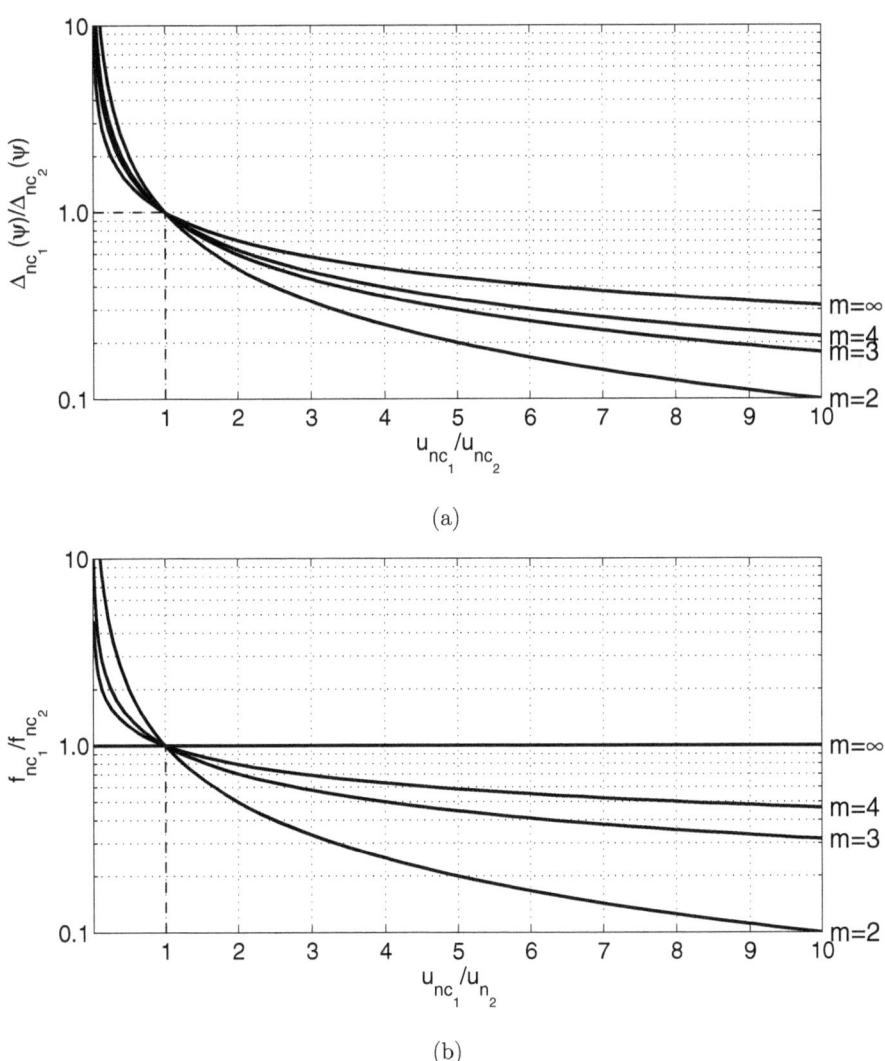

Figure 3.4: Impact of m on the ratio of $\Delta_{nc}(\psi)$ and f_{nc} values for different values of m. (a) Ratio between $\Delta_{nc}(\psi)$ values. (b) Ratio between f_{nc} values.

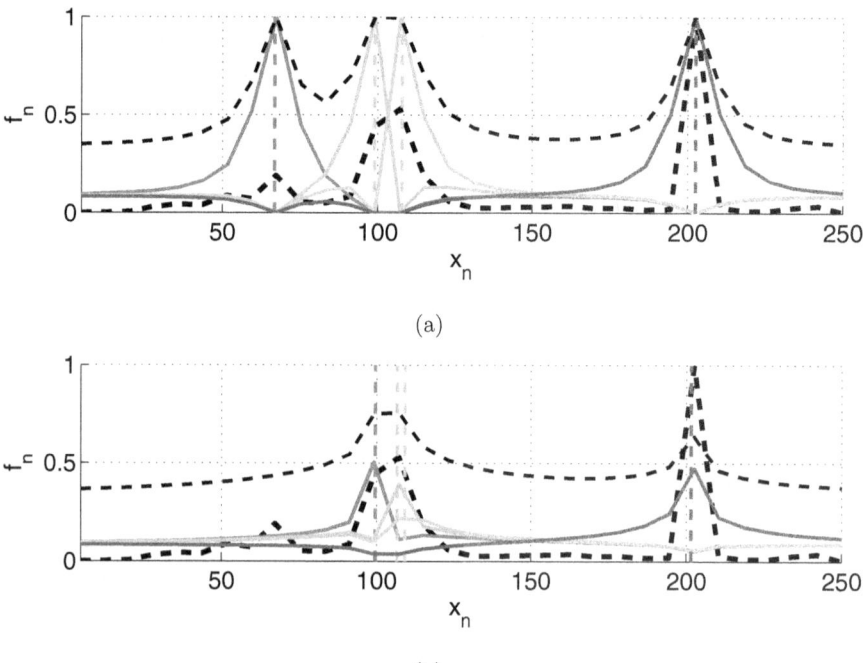

Figure 3.5: Impact of m on the fuzziness of the output of the clustering algorithm. (a) $m = 2$. (b) $m = 5$.

Chapter 4

Experimental Results

The algorithm developed in this work is implemented as the Matlab class *Ava*. This class utilizes the virtual function *phi()* which it allows child classes to override according to the specifications of different problem classes. These child classes also perform the load and visualization operations relevant to the cluster and data item models defined by the respective problem classes. In this work, we utilize six such child classes. The problem classes handled by these child classes are listed in Table 4.1. We emphasize that the child classes do not contain any piece of the method developed in this work.

Figure 4.1 shows the inputs and outputs of a *2dl* problem instance. This problem instance is concerned with finding lines in a set of weighted 2D data points. Here, Figure 4.1(a)-top shows the set of 593 input data items. In this graph, every data item is represented as a circle and the shades of gray and the sizes of the circles denote the values of ω_n. In this experiment, we set $C = 4$ and $m = 3$ and then apply the developed algorithm. This process starts with the cluster seeds shown in Figure 4.1(b). After 27 iterations and a total of 719 milliseconds the process converges. Figure 4.1(c) shows the approach of Δ towards the value that it settles on. This period is separated into two sections. First, m is set to 3, and when change in Δ becomes negligible, m is dropped to 2. These two periods are shown in the top chart in Figure 4.1(c). Following this process, f_{nc} and p_n converge to the histograms shown in Figures 4.1(c)-middle and -bottom. Note that, f_{nc} and p_n are in fact only theoretically present in the process and are never calculated.

Figure 4.1(d) shows the clusters which the developed algorithm converges to. This set of clusters are then utilized in order to generate the classification depicted in Figure 4.1(e). Here, each color denotes one cluster and black represents the outliers.

Table 4.1: Properties of the problem classes utilized in this paper.

Problem Class	x_n	ψ_c	Purpose	$\phi(x_n, \psi_c)$
2dc	$x_n \in \mathbb{R}^2$	$\psi_c = [m_c, \rho_c]$ $m_c \in \mathbb{R}^2$ $\rho_c > 0$	Finding Circles	$\left[\|x_n - m_c\|^2 - \rho_c^2 \right]^2$
2de	$x_n \in \mathbb{R}^2$	$\psi_c \in \mathbb{R}^2$	Euclidean Clustering	$\|x_n - \psi_c\|^2$
2dl	$x_n \in \mathbb{R}^2$	$\psi_c = [m_c, v_c]$ $m_c \in \mathbb{R}^2$ $v_c \in \mathbb{R}^2, \|v_c\| = 1$	Finding Lines	$\|x_n - m_c - v_c^T (x_n - m_c) v_c\|^2$
3dpp	$x_n \in \mathbb{R}^3$	$\psi_c \in \mathbb{R}^3$	Finding Planes	$\frac{1}{\|\psi_c\|^2} \left(\psi_c^T x_n - \|\psi_c\|^2 \right)^2$
ics	$x_n \in \mathbb{R}^3$	$\psi_c = [m_c, v_c]$ $m_c \in \mathbb{R}^3$ $v_c \in \mathbb{R}^3, \|v_c\| = 1$	Color Image Segmentation	$\|x_n - m_c - v_c^T (x_n - m_c) v_c\|^2$
ighe	$x_n \in \mathbb{R}$	$\psi_c \in \mathbb{R}$	Grayscale Image Segmentation	$(x_n - \psi_c)^2$

We note that the developed method is governed by the three configuration parameters m, C, and φ. As will be discussed later, we choose the value of $m = 3$ and rely on an external process in order to recommend the value of C. The third side of this triangle, i.e. φ, is set according to the process outlined in Section 3.5. Here, we review the significance of utilizing proper values for these configuration variables.

Figure 4.2 shows the impact of the value of m on the converged clusters for a *3dpp* problem instance. This problem class is concerned with finding planar sections in range data captured by a *Kinect 2* sensor. The depth-maps used in this experiment are captured at the resolution of 424×512 pixels. Here, intrinsic parameters of the camera are acquired through the Kinect SDK and each data item in this problem class has the weight of one.

The data shown in Figure 4.2 corresponds to a room in which three human bodies are present. As seen in this figure, this set of input data contains three planes, i.e. two walls and the floor. Therefore, in this context, the aim of the *3dpp* problem class is to recognize these three planes and to reject the data items which correspond to the human bodies as outliers.

Note that the value of m carried in the caption of Figure 4.2 in fact corresponds to the starting

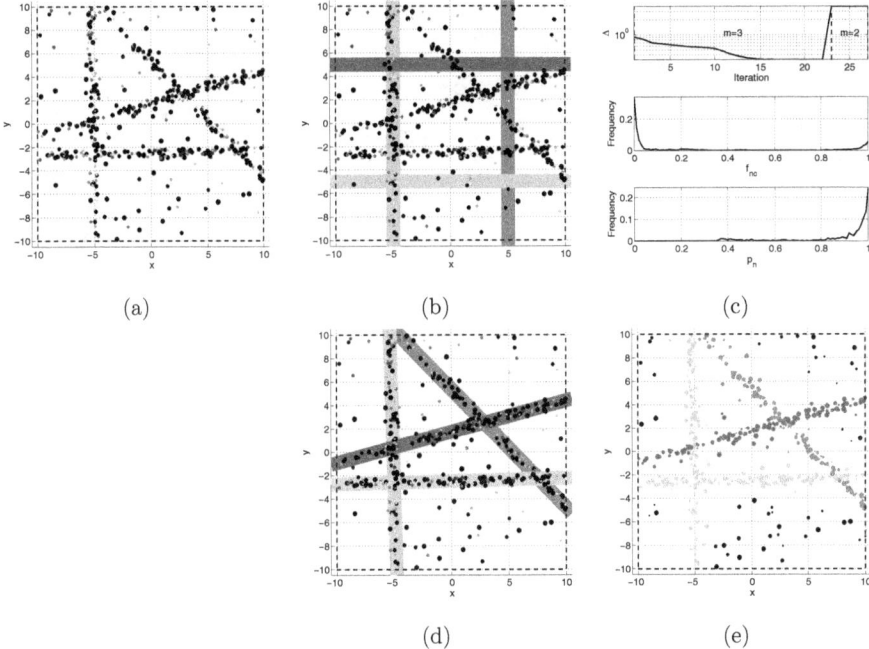

Figure 4.1: Inputs and outputs of the developed algorithm for a *2dl* problem instance. (a) Input data items. (b) Cluster seeds. (c) Internal variables. (d) Converged clusters. (e) Classification results.

value of m. As discussed in Section 3.6, m can start at different values, but, nevertheless, it is always lowered to the value of 2, unless explicitly specified differently.

It is informative to compare the converged planes in Figure 4.2, particularly between Figures 4.2(a) and (c), and especially for the wall to the left of the camera. In fact, we observe in Figure 4.2(a) that the plane corresponding to this particular wall does not converge to it properly. This situation is more evident when we compare the position of the same wall between Figures 4.2(a) and (b). In other words, the solution depicted in Figure 4.2(a) has not in fact converged to the global minimum. We argue that the smaller m used in Figure 4.2(a), compared to Figure 4.2(b), does not "smooth" the local minimums in the cost manifold, and, therefore, the optimization procedure fails to converge to the desired minimum.

The situation present in Figure 4.2(c) is in fact the other end of the spectrum. In this case, it is observed that m is "too large", and that the three clusters have converged to a subset of data items

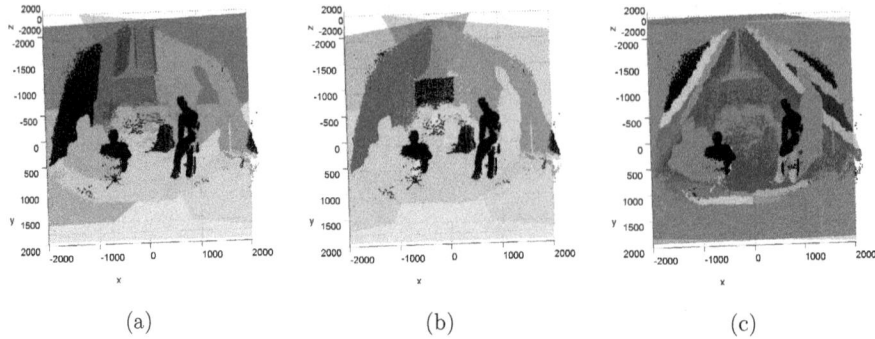

Figure 4.2: Impact of the choice of m on the converged clusters for a *3dpp* problem instance. (a) $m = 2$. (b) $m = 3$. (c) $m = 4$. Input data courtesy of *Epson Edge*.

which contains elements from the three planes as well as sections which belong to the individuals present in the scene. We emphasize that this is a local minimum situation, in which the clusters fail to "recognize" the planar sections of the data. We conjecture that this phenomenon happens when m is too large and hence the manifold on which the optimization process occurs is excessively smoothed. Under such conditions, i.e. when m is too large, the clusters move to undesired positions in the search space and cannot leave them into more optimal positions when m is lowered.

We emphasize that the condition depicted in Figure 4.2 is an exceptional situation which is found trough trial-and-error and that the value of m does not carry such a significant influence. Nevertheless, the proper choice of m is a compromise between an under-smoothed optimization manifold which contains numerous local minimums that trap the solution and an over-smoothed manifold which eliminates the preferred minimum. As will be demonstrated later, $m = 3$ appears to be an optimal compromise in this context.

The input data used in the experiments shown in Figure 4.2 contains 45,169 data items. The developed algorithm converges after 40, 39, and 221 iterations, for $m = 2$, $m = 3$, and $m = 4$, respectively. For the specified settings, the developed algorithm takes 5,448, 6,130, and 47,139 milliseconds to converge, respectively.

Figure 4.3 shows the results corresponding to a *2dc* problem instance for different values of C. Here, the input contains 805 data items which are generated using the superposition of 3 noise-contaminated circles plus additional outlier points.

As seen in Figures 4.3(a), when the proposed algorithm seeks a smaller number of clusters, than

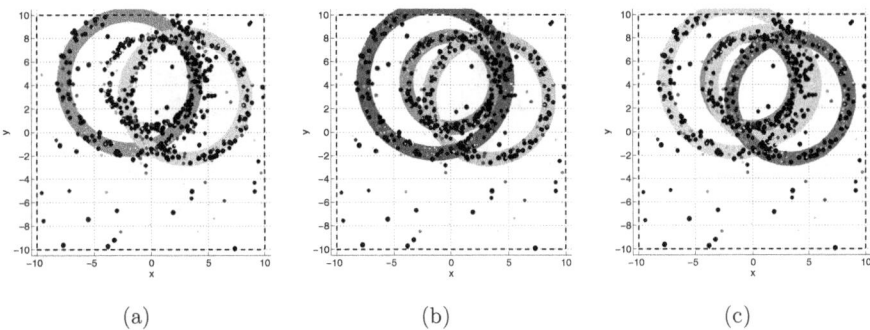

Figure 4.3: Impact of the choice of C on the converged clusters for a *2dc* problem instance. (a) $C = 2$. (b) $C = 3$. (c) $C = 4$.

what is available in the data, it converges to C clusters and labels the rest of the data as outliers. In the case of Figure 4.3(c), however, C is greater than the actual number of clusters present in the data. Hence, the algorithm "recognizes" an additional cluster which is in effect composed of pieces of others clusters as well as outliers. In contrast, the case of Figure 4.3(b) corresponds to a situation in which C matches the actual number of clusters.

The situation outlined above is an inherent aspect of the wide range of fuzzy clustering algorithms. In essence, the proposed algorithm relies on the assumption that a proper estimate for the number of clusters is known. It is important to emphasize, however, that when C is selected incorrectly, either smaller or greater than required, the proposed algorithm converges to useful clusters. Nevertheless, an exogenous means of determining the proper value of C is a beneficial prefix to the proposed method.

As discussed in Section 2.4, VAT algorithms are specifically designed for the purpose of suggesting an appropriate value for C without actually performing the clustering process. It is important to reiterate, however, that a majority of the VAT methods available in the literature are defined in a Euclidean context. In fact, to the best of our knowledge, a generic class-independent VAT method does not exist. Research into the generalization of the VAT framework is, to our understanding, a worthy direction for work.

The results carried in Figure 4.3 are generated after 31, 29, and 35 iterations, for $C = 2$, $C = 3$, and $C = 4$, respectively. In these settings, the process takes 715, 891, and 1,579 milliseconds to complete, respectively.

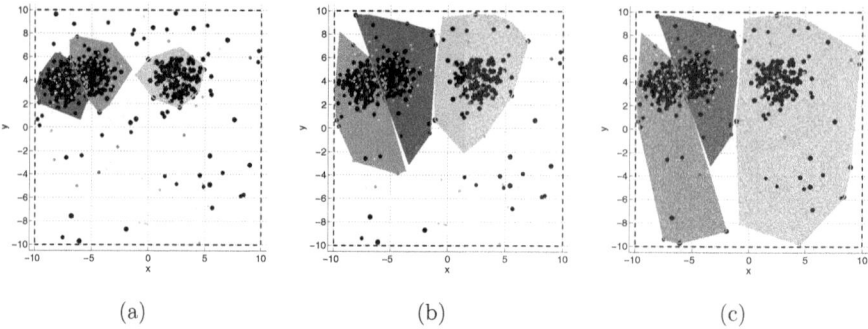

Figure 4.4: Impact of the choice of φ on the converged clusters for a *2de* problem instance. (a) $\varphi = \frac{1}{4}\varphi_o$. (b) $\varphi = \varphi_o$. (c) $\varphi = 4\varphi_o$.

Figure 4.4 shows three sets of results which correspond to a *2de* problem instance. This problem class is concerned with finding compact clusters in weighted points on a 2D plane. Here, we have used the value of φ_o which corresponds to the theoretically appropriate value of φ, based on the parameters of the data generation procedure. In this experiment we examine the effects of choosing φ_o as well as under– and overestimating the value of φ by a factor of 4.

As seen in Figure 4.4(a), an underestimated φ shrinks the clusters. In essence, this situation can be explained as both f_{nc} and p_n dropping faster as ϕ_{nc} increases. Similarly, an overestimated φ bloats the clusters, for the same reason, as seen in Figure 4.4(c).

There are 849 data items in the input utilized for the experiments presented in Figure 4.4. These results are generated after 36, 36, and 37 iterations. It takes the proposed algorithm 2,361, 2,114, and 1,862 milliseconds to converge for the results carried in Figure 4.4.

We note that, the value of φ does not have a direct impact on the number of iterations required by the proposed algorithm before it converges. This is contrary to the impact of m and C on the execution time of the developed algorithm. We note that when m increases, the developed algorithm has to spend additional number of iterations in order for m to decrease to its final value. An increase in C inflates the execution time of the developed algorithm for a different reason. While it may take longer for the algorithm to converge when C increases, a higher C in fact directly increases the cost of every iteration.

We conclude this section with additional problem instances from the six problem classes listed in Table 4.1.

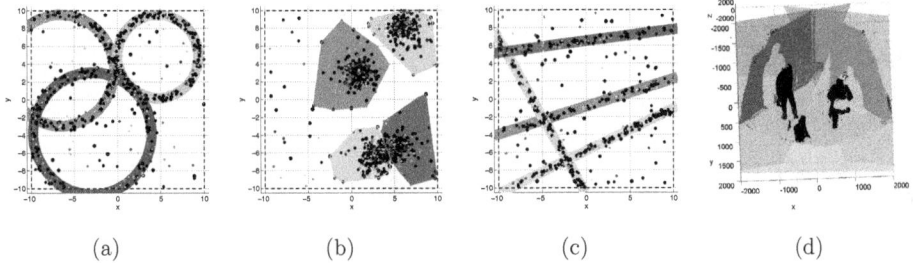

(a) (b) (c) (d)

Figure 4.5: Results corresponding to the problem classes listed in Table 4.1. (a) *2dc*. (b) *2de*. (c) *2dl*. (d) *3dpp*.

Figure 4.5(a) carries the output generated by the developed algorithm for a *2dc* problem instance. Here, we have used $m = 3$ and $C = 3$. The developed algorithm in this case converges after 45 iterations and 1,487 milliseconds of operation. This problem instance contains 712 data items. Similarly, the visualization carried in Figure 4.5(b) exhibits the output generated by the developed algorithm for a *2de* problem instance which contains 1,096 data items. Here, we utilize $m = 3$ and $C = 4$ and the algorithm converges after 49 iterations and 4,004 milliseconds of calculation.

Figure 4.5(c) corresponds to a *2dl* problem instance which contains 567 data items. We utilize $m = 3$ and $C = 4$ and the developed algorithm converges after 86 iterations and 2,067 milliseconds of operation. Finally, Figure 4.5(d) carries the solution generated by the developed algorithm for a *3dpp* problem instance with 43,315 data items. Here, we set $m = 3$ and $C = 3$ and the developed algorithm converges after 33 iterations and 9,515 milliseconds of operation.

Figure 4.6 carries the results generated for an *ics* problem instance. Here, Figure 4.6(a) shows the input set of data items and Figure 4.6(b) exhibits the clusters generated by the developed algorithms. These clusters result in the classification carried in Figure 4.6(c). Here, we have used $m = 3$ and $C = 3$. The developed algorithm converges after 68 iterations and 29,752 milliseconds of calculation.

Figure 4.7 contains the results of applying the developed algorithm on an *ighe* problem instance. Here, Figure 4.7(a) carries the input set of data items, i.e. the standard image *Cameraman*. The developed algorithm produces the clusters shown in Figure 4.7(b). These clusters result in the classification shown in Figure 4.7(c). Here, we have used $m = 3$ and $C = 2$. The developed algorithm converges for this problem instance after 30 iterations and 471 milliseconds of calculation.

We note that we utilized $m = 3$ for every one of the six problem instances carried in Figures 4.5,

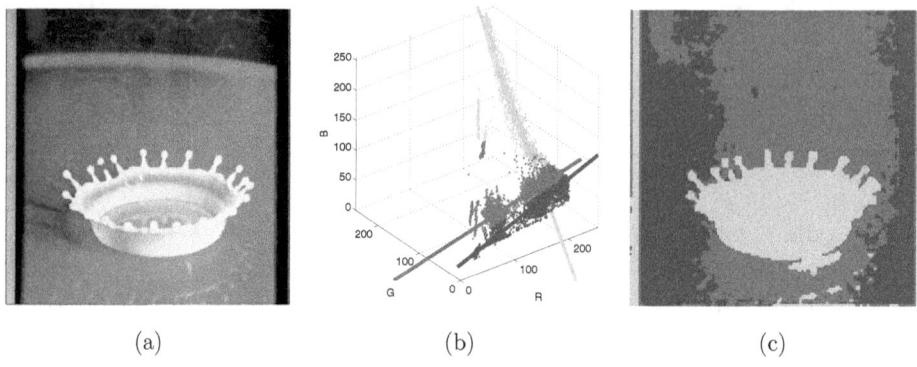

Figure 4.6: Results corresponding to an *ics* problem instance. (a) Input data items. (b) Converged clusters. (c) Classification results.

4.6, and 4.7. Moreover, in every other experiment carried in this paper we used $m = 3$, unless the purpose of the experiment was to investigate the impact of other values of m. We conjecture that the "smoothing" of the optimization manifold achieved through setting $m = 3$ is at the desired level for many problem classes. This statement, obviously, may not apply to any arbitrary problem class. Nevertheless, we argue that while the actual value of m needs to be adjusted with some level of care, $m = 3$ is an acceptable option, at least for the six problem classes reviewed in this paper.

In effect, the outcome of the developed algorithm is affected by the values set for the three configuration parameters m, C, φ. As discussed in the previous paragraph, there exists a default value for m. In the event that this value of m is not appropriate for a particular problem class, thus resulting in the developed algorithm failing to converge to an acceptable minimum, we emphasize that the proper value of m is not problem instance dependent. In other words, if $m = 3$ is not appropriate for a new problem class, one needs to set it at the level of problem class, and not problem instances. Setting the value of C is a more important consideration, though. It is known in the literature that improper C can result in portions of the data being classified as outlier or multiple clusters converging to the same sections of the input data items. While the category of VAT algorithms is precisely designed to tackle this problem, VATs are commonly Euclidean. Hence, the development of a generic non-Euclidean VAT is a helpful prefix for the developed algorithm and will make it more suitable for usage in applications.

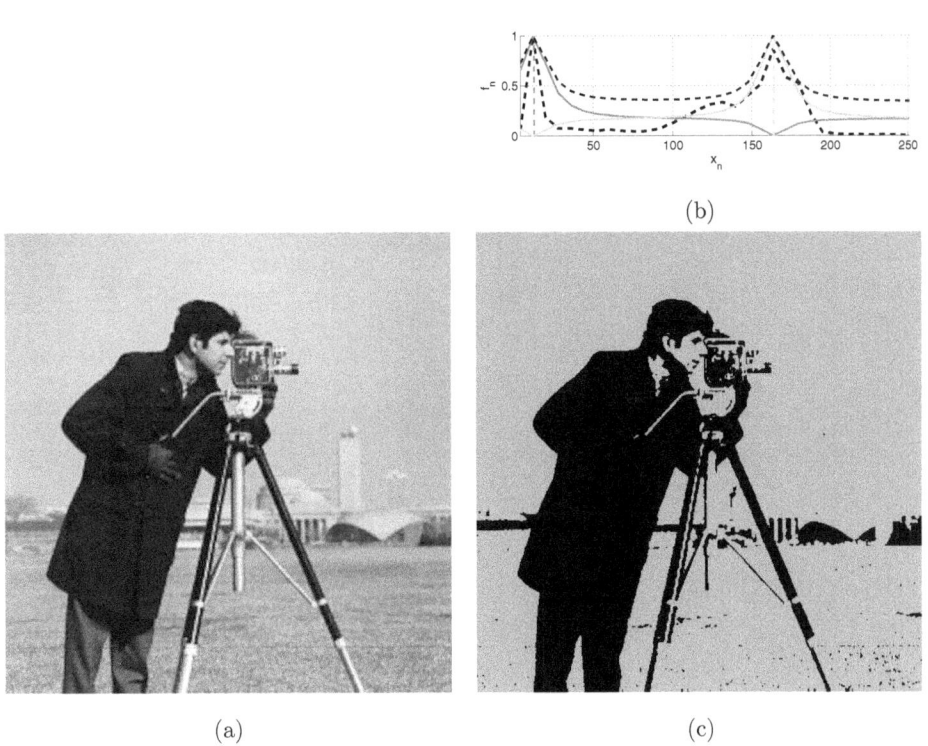

Figure 4.7: Results corresponding to an *ighe* problem instance. (a) Input data items. (b) Converged clusters. (c) Classification results.

Chapter 5

Conclusions

In this paper, we proposed a novel algorithm for performing unsupervised fuzzy clustering within a class-independent framework. This framework utilizes the data items and the clusters as abstract mathematical objects. These entities are assumed to be related together through a given data item-to-cluster distance function. Hence, the core of the developed algorithm is independent of the particularities of the data item and cluster models as well as the particular distance function relevant to them. We show that the loss model for this system can be written as a least-square cost function and demonstrate a minimization strategy for it which utilizes Picard iterations of the Levenberg-Marquardt optimization algorithm. We provide a framework of analysis for the different choices for the fuzzifier and show that the step from HCM to FCM can be extended in order to provide assurance of convergence for the developed algorithm. We provide construction procedures for the two parameters in the system which are to be set for every problem class. As a result, we exhibit that the developed method is independent of any user supervision for any particular problem instance. The paper contains experimental results collected for six different problem classes. We note that, the applicability of the results generated by the developed method are further enhanced when the number of clusters present in the data is known *a priori*. We envision that work on a generic non-Euclidean VAT algorithm could be a step in this direction.

Acknowledgments

We wish to thank the management of *Intellijoint Surgical Inc.* for their support. We thank the management of *Epson Edge, Epson Canada Limited* for allowing us to use the Kinect data utilized in some of the experiments carried in this paper. The idea for this paper was conceived after watching *Ex Machina* by the British director *Alex Garland*.

Bibliography

[1] J. B. MacQueen, Some methods for classification and analysis of multivariate observations, in: Proceedings of 5-th Berkeley Symposium on Mathematical Statistics and Probability, Berkeley, 1967, pp. 281–297.

[2] R. Gray, Y. Linde, Vector quantizers and predictive quantizers for Gauss-Markov sources, IEEE Transactions on Communications 30 (2) (1982) 381–389.

[3] T. Kanungo, D. M. Mount, N. S. Netanyahu, C. D. Piatko, R. Silverman, A. Y. Wu, An efficient k-means clustering algorithm: Analysis and implementation, IEEE Transactions on Pattern Analysis and Machine Intelligence 24 (7) (2002) 881–892.

[4] G. H. Ball, D. J. Hall, A clustering technique for summarizing multivariate data, Behavioral Science 12 (2) (1967) 153–155.

[5] L. A. Zadeh, Fuzzy sets, Information Control 8 (1965) 338–353.

[6] M.-S. Yang, A survey of fuzzy clustering, Mathematical and Computer Modelling 18 (11) (1993) 1–16.

[7] A. Baraldi, P. Blonda, A survey of fuzzy clustering algorithms for pattern recognition. I & II, IEEE Transactions on Systems, Man, and Cybernetics, Part B: Cybernetics 29 (6) (1999) 778–801.

[8] B. Abu-Jamous, R. Fa, A. K. Nandi, Fuzzy Clustering, John Wiley & Sons, Ltd, 2015, pp. 167–179.

[9] J. C. Bezdek, J. Dunn, Optimal fuzzy partitions: A heuristic for estimating the parameters in a mixture of normal distributions, IEEE Transactions on Computers C-24 (8) (1975) 835–838.

[10] S. Miyamoto, Fuzzy clustering - Basic ideas and overview, in: J. Kacprzyk, W. Pedrycz (Eds.), Springer Handbook of Computational Intelligence, Springer Berlin Heidelberg, 2015, pp. 239–248.

[11] E. H. Ruspini, A new approach to clustering, Information & Control 15 (1) (1969) 22–32.

[12] E. H. Ruspini, Numerical methods for fuzzy clustering, Information Sciences 2 (3) (1970) 319–350.

[13] J. C. Dunn, A fuzzy relative of the ISODATA process and its use in detecting compact well-separated clusters, Journal of Cybernetics 3 (3) (1973) 32–57.

[14] J. C. Dunn, Well separated clusters and optimal fuzzy partitions, Journal of Cybernetics 4 (1974) 95–104.

[15] J. C. Bezdek, Fuzzy mathematics in pattern classification, Ph.D. thesis, Cornell University (1973).

[16] J. C. Bezdek, Pattern Recognition with Fuzzy Objective Function Algorithms, Plenum Press, New York, 1981.

[17] J. M. Leski, Generalized weighted conditional fuzzy clustering, IEEE Transactions on Fuzzy Systems 11 (6) (2003) 709–715.

[18] J. Yu, Q. Cheng, H. Huang, Analysis of the weighting exponent in the FCM, IEEE Transactions on Systems, Man, and Cybernetics, Part B: Cybernetics 34 (1) (2004) 634–639.

[19] M. Trivedi, J. C. Bezdek, Low-level segmentation of aerial images with fuzzy clustering, IEEE Transactions on Systems, Man, and Cybernetics 16 (4) (1986) 589–598.

[20] H. Frigui, R. Krishnapuram, A robust algorithm for automatic extraction of an unknown number of clusters from noisy data, Pattern Recognition Letters 17 (12) (1996) 1223–1232.

[21] F. Klawonn, R. Kruse, H. Timm, Fuzzy shell cluster analysis, in: G. della Riccia, H. Lenz, R. Kruse (Eds.), Learning, networks and statistics, Springer, 1997, pp. 105–120.

[22] J. C. Bezdek, A physical interpretation of fuzzy ISODATA, IEEE Transactions on Systems, Man and Cybernetics SMC-6 (5) (1976) 387–389.

[23] N. R. Pal, J. C. Bezdek, On cluster validity for the fuzzy C-means model, IEEE Transactions on Fuzzy Systems 3 (3) (1995) 370–379.

[24] C. Borgelt, Objective functions for fuzzy clustering, in: C. Moewes, A. Nurnberger (Eds.), Computational Intelligence in Intelligent Data Analysis, Vol. 445 of Studies in Computational Intelligence, Springer Berlin Heidelberg, 2013, pp. 3–16.

[25] F. Klawonn, F. Hoppner, What is fuzzy about fuzzy clustering? Understanding and improving the concept of the fuzzifier, in: M. R. Berthold, H.-J. Lenz, E. Bradley, R. Kruse, C. Borgelt (Eds.), Advances in Intelligent Data Analysis V, Vol. 2810 of Lecture Notes in Computer Science, Springer Berlin Heidelberg, 2003, pp. 254–264.

[26] I. H. Suh, J.-H. Kim, F. Chung-Hoon Rhee, Convex-set-based fuzzy clustering, IEEE Transactions on Fuzzy Systems 7 (3) (1999) 271–285.

[27] R. Kruse, C. Doring, M.-J. Lesot, Fundamentals of fuzzy clustering, in: J. V. de Oliveira, W. Pedrycz (Eds.), Advances in Fuzzy Clustering and its Applications, Wiley, England, 2007, pp. 3–29.

[28] R. Yager, D. Filev, Approximate clustering via the mountain method, IEEE Transactions on Systems, Man and Cybernetics 24 (8) (1994) 1279–1284.

[29] J. M. Leski, Fuzzy c-varieties/elliptotypes clustering in reproducing kernel Hilbert space, Fuzzy Sets and Systems 141 (2) (2004) 259–280.

[30] C. Borgelt, C. Braune, M.-J. Lesot, R. Kruse, Handling noise and outliers in fuzzy clustering, in: D. E. Tamir, N. D. Rishe, A. Kandel (Eds.), Fifty Years of Fuzzy Logic and its Applications, Vol. 326 of Studies in Fuzziness and Soft Computing, Springer International Publishing, 2015, pp. 315–335.

[31] S. Miyamoto, D. Suizu, Fuzzy c-means clustering using kernel functions in support vector machines, Journal of Advanced Computational Intelligence and Intelligent Informatics 7 (1) (2003) 25–30.

[32] D.-M. Tsai, C.-C. Lin, Fuzzy C-means based clustering for linearly and nonlinearly separable data, Pattern Recognition 44 (8) (2011) 1750–1760.

[33] K.-L. Wu, M.-S. Yang, Alternative C-means clustering algorithms, Pattern Recognition 35 (10) (2002) 2267–2278.

[34] L. Kaufman, P. J. Rousseeuw, Finding Groups in Data: an Introduction to Cluster Analysis, John Wiley & Sons Inc, New York, 1990.

[35] R. J. Hathaway, J. W. Davenport, J. C. Bezdek, Relational duals of the C-means clustering algorithms, Pattern Recognition 22 (2) (1989) 205–212.

[36] R. J. Hathaway, J. C. Bezdek, NERF C-means: Non-Euclidean relational fuzzy clustering, Pattern Recognition 27 (3) (1994) 429–437.

[37] L. Fu, E. Medico, FLAME, A novel fuzzy clustering method for the analysis of DNA microarray data, BMC Bioinformatics 8 (3).

[38] T. Hastie, R. Tibshirani, J. Friedman, The Elements of Statistical Learning, Springer, New York, 2009.

[39] K. Jajuga, L_1-norm based fuzzy clustering, Fuzzy Sets and Systems 39 (1) (1991) 43–50.

[40] L. Bobrowski, J. C. Bezdek, C-means clustering with the ℓ_1 and ℓ_∞ norms, IEEE Transactions on Systems, Man, and Cybernetics 21 (3) (1991) 545–554.

[41] R. J. Hathaway, J. C. Bezdek, Y. Hu, Generalized fuzzy C-means clustering strategies using L_p norm distances, IEEE Transactions on Fuzzy Systems 8 (5) (2000) 576–582.

[42] N. B. Karayiannisa, M. M. Randolph-Gips, Non-Euclidean C-means clustering algorithms, Intelligent Data Analysis 7 (2003) 405–425.

[43] D. E. Gustafson, W. C. Kessel, Fuzzy clustering with a fuzzy covariance matrix, in: IEEE Conference on Decision and Control including the 17th Symposium on Adaptive Processes, Vol. 17, San Diego, CA, 1979, pp. 761–766.

[44] I. Gath, A. Geva, Unsupervised optimal fuzzy clustering, IEEE Transaction on Pattern Analysis Machine Intelligence 11 (7) (1989) 773–781.

[45] R. J. Hathaway, J. C. Bezdek, Switching regression models and fuzzy clustering, IEEE Transactions on Fuzzy Systems 1 (3) (1993) 195–204.

[46] H. Frigui, R. Krishnapuram, A comparison of fuzzy shell-clustering methods for the detection of ellipses, IEEE Transactions on Fuzzy Systems 4 (2) (1996) 193–199.

[47] R. N. Dave, R. Krishnapuram, Robust clustering methods: A unified view, IEEE Transactions on Fuzzy Systems 5 (2) (1997) 270–293.

[48] K. K. Chintalapudi, M. Kam, The credibilistic fuzzy C-means clustering algorithm, in: IEEE International Conference on Systems, Man, and Cybernetics (SMC 1998), Vol. 2, 1998, pp. 2034–2039.

[49] N. R. Pal, K. Pal, J. M. Keller, J. C. Bezdek, A possibilistic fuzzy c-means clustering algorithm, IEEE Transactions on Fuzzy Systems 13 (4) (2005) 517–530.

[50] J. Leski, Towards a robust fuzzy clustering, Fuzzy Sets and Systems 137 (2) (2003) 215–233.

[51] P. D'Urso, L. D. Giovanni, Robust clustering of imprecise data, Chemometrics and Intelligent Laboratory Systems 136 (2014) 58–80.

[52] J. J. D. Gruijter, A. B. McBratney, A modified fuzzy K-means method for predictive classification, in: H. H. Bock (Ed.), Classification and Related Methods of Data Analysis, Elsevier, Amsterdam, The Netherlands, 1988, pp. 97–104.

[53] R. N. Dave, Characterization and detection of noise in clustering, Pattern Recognition Letters 12 (11) (1991) 657–664.

[54] Y. Ohashi, Fuzzy clustering and robust estimation, Presented at the 9th SAS Users Group International (SUGI) Meeting at Hollywood Beach, Florida. (1984).

[55] R. N. Dave, Robust fuzzy clustering algorithms, in: Second IEEE International Conference on Fuzzy Systems, Vol. 2, 1993, pp. 1281–1286.

[56] R. Krishnapuram, J. M. Keller, A possibilistic approach to clustering, IEEE Transactions on Fuzzy Systems 1 (2) (1993) 98–110.

[57] M. Barni, V. Cappellini, A. Mecocci, Comments on "A possibilistic approach to clustering", IEEE Transactions on Fuzzy Systems 4 (3) (1996) 393–396.

[58] H. Timm, C. Borgelt, C. Doring, R. Kruse, An extension to possibilistic fuzzy cluster analysis, Fuzzy Sets and Systems 147 (1) (2004) 3–16.

[59] N. R. Pal, K. Pal, J. C. Bezdek, A mixed c-means clustering model, in: Proccedings of the Sixth IEEE International Conference on Fuzzy Systems, Vol. 1, 1997, pp. 11–21.

[60] R. Dave, S. Sen, On generalising the noise clustering algorithms, in: Proceedings of the 7th IFSA World Congress (IFSA 1997), 1997, pp. 205–210.

[61] N. R. Pal, K. Pal, J. M. Keller, J. C. Bezdek, A new hybrid C-means clustering model, in: Proceedings of the 2004 IEEE International Conference on Fuzzy Systems, Vol. 1, 2004, pp. 179–184.

[62] J.-S. Zhang, Y.-W. Leung, Improved possibilistic C-means clustering algorithms, IEEE Transactions on Fuzzy Systems 12 (2) (2004) 209–217.

[63] R. Krishnapuram, J. M. Keller, The possibilistic C-means algorithms: Insights and recommendations, IEEE Transactions on Fuzzy Systems 4 (3) (1996) 385–393.

[64] F. Masulli, S. Rovetta, Soft transition from probabilistic to possibilistic fuzzy clustering, IEEE Transactions on Fuzzy Systems 14 (4) (2006) 516–527.

[65] A. Abadpour, Rederivation of the fuzzypossibilistic clustering objective function through Bayesian inference, Fuzzy Sets and Systems 305 (2016) 29–53.

[66] A. Abadpour, A sequential bayesian alternative to the classical parallel fuzzy clustering model, Information Sciences 318 (2015) 28–47.

[67] R. Krishnapuram, C.-P. Freg, Fitting an unknown number of lines and planes to image data through compatible cluster merging, Pattern Recognition 25 (4) (1992) 385–400.

[68] R. Krishnapuram, H. Frigui, O. Nasraoui, Fuzzy and possibilistic shell clustering algorithms and their application to boundary detection and surface approximation - Parts I & II, IEEE Transaction on Fuzzy Systems 3 (1) (1995) 29–60.

[69] J. M. Jolion, P. Meer, S. Bataouche, Robust clustering with applications in computer vision, IEEE Transactions on Pattern Analysis and Machine Intelligence 13 (8) (1991) 791–802.

[70] J. C. Bezdek, N. R. Pal, Some new indexes of cluster validity, IEEE Transactions on Systems, Man, and Cybernetics, Part B: Cybernetics 28 (3) (1998) 301–315.

K. Rose, E. Gurewitz, G. C. Fox, Vector quantization by deterministic annealing, IEEE Transactions on Information Theory 38 (4) (1992) 1249–1257.

G. Beni, X. Liu, A least biased fuzzy clustering method, IEEE Transactions on Pattern Analysis and Machine Intelligence 16 (9) (1994) 954–960.

J. J. More, The Levenberg-Marquardt algorithm: Implementation and theory, in: G. Watson (Ed.), Numerical Analysis, Vol. 630 of Lecture Notes in Mathematics, Springer Berlin Heidelberg, 1978, pp. 105–116.

K. Levenberg, A method for the solution of certain non-linear problems in least squares, Quarterly Journal of Applied Mathmatics II (2) (1944) 164–168.

D. W. Marquardt, An algorithm for least-squares estimation of nonlinear parameters, Journal of the Society for Industrial and Applied Mathematics 11 (2) (1963) 431–441.

[71] I. Sledge, J. C. Bezdek, T. C. Havens, J. M. Keller, Relational generalizations of clust indices, IEEE Transactions on Fuzzy Systems 18 (4) (2010) 771–786.

[72] J. C. Bezdek, R. J. Hathaway, VAT: A tool for visual assessment of (cluster) ter Proceedings of the 2002 International Joint Conference on Neural Networks (IJC Vol. 3, 2002, pp. 2225–2230.

[73] L. Wang, U. T. Nguyen, J. C. Bezdek, C. A. Leckie, K. Ramamohanarao, iVAT a Enhanced visual analysis for cluster tendency assessment, in: M. J. Zaki, J. X. Yu, dran, V. Pudi (Eds.), Advances in Knowledge Discovery and Data Mining, Vol. 6118 Notes in Computer Science, Springer Berlin Heidelberg, 2010, pp. 16–27.

[74] T. C. Havens, J. C. Bezdek, An efficient formulation of the improved visual ass cluster tendency (iVAT) algorithm, IEEE Transactions on Knowledge and Data E 24 (5) (2012) 813–822.

[75] N. Metropolis, A. W. Rosenbluth, M. N. Rosenbluth, A. H. Teller, E. Teller, Equat calculations by fast computing machines, Journal of Chemical Physics 21 (6) (1953)

[76] S. Kirkpatrick, C. D. Gelatt, J. M. P. Vecchi, Optimization by simulated anneali 220 (4598) (1983) 671–680.

[77] K. Rose, Deterministic annealing for clustering, compression, classification, regr related optimization problems, Proceedings of the IEEE 86 (11) (1998) 2210–2239

[78] R. Durbin, R. Szeliski, A. Yuille, An analysis of the elastic net approach to the Salesman Problem, Neural Computation 1 (3) (1989) 348–358.

[79] P. D. Simic, Statistical mechanics as the underlying theory of "elastic" and "neur sations, Network: Computation in Neural Systems 1 (1) (1990) 89–103.

[80] D. Geiger, F. Girosi, Parallel and deterministic algorithms from MRFs: surface reco IEEE Transactions on Pattern Analysis and Machine Intelligence 13 (5) (1991) 40

[81] K. Rose, E. Gurewitz, G. Fox, A deterministic annealing approach to clusterin Recognition Letters 11 (9) (1990) 589–594.

[82] K. Rose, E. Gurewitz, G. C. Fox, Statistical mechanics and phase transitions in Physical Review Letters 65 (1990) 945–948.